Cannonballs made of iron

Iron

Peter Murray

21.36
SA

A^{+}

Smart Apple Media

COPYRIGHT

☼ Published by Smart Apple Media

1980 Lookout Drive, North Mankato, MN 56003

Designed by Rita Marshall

Copyright © 2002 Smart Apple Media. International copyright reserved in all countries. No part of this book may be reproduced in any form without written permission from the publisher.

Printed in the United States of America

☼ Photographs by Pat Berrett, Richard Cummins, JLM Visuals (Richard Jacobs), Bruce Leighty, Photri (R. Harding), Rainbow (Jim McGrath), Unicorn Stock Photos (Nancy Ferguson)

☼ Library of Congress Cataloging-in-Publication Data

Murray, Peter. Iron / by Peter Murray. p. cm. – (From the earth)

Includes index.

☼ ISBN 1-58340-113-X

1. Iron–Juvenile literature. [1. Iron.] I. Title. II. Series.

TN705 .M87 2001   669.141–dc21   00-068796

☼ First Edition 9 8 7 6 5 4 3 2 1

Iron

CONTENTS

Iron from Above

In 1922, the 3,200-year-old tomb of King Tutankhamen was discovered in Egypt. The tomb was filled with beautiful objects made of gold, silver, and jewels. There was more gold in Tutankhamen's tomb than in the Royal Bank of Egypt! ☀

One of the rarest treasures found in the tomb was a dagger. The dagger had a solid gold handle decorated with crystals, and a sheath made of gold—but what made it special was the hard, gray, metal blade. When Tutankhamen ruled Egypt, that

The gold mask of King Tutankhamen

gray metal was more valuable than diamonds or gold. King

Tut's dagger was made of iron. ☀ Iron is the fourth

most common **element** on Earth. The earth's

crust is more than five percent iron. Iron is 700

times more common than copper and 12 billion

times more common than gold! But 3,000 years

ago, iron was the rarest of metals. People did not

Some meteorites are solid iron and nickel, a natural form of stainless steel.

know that the earth contained a fortune in iron. The only iron

they knew came from rare, oddly shaped rocks. ☀ We now

Varieties of iron ore

know that these iron rocks came from outer space. King Tut's

dagger may have been made from part of a meteorite.

The Iron Age

The first iron used by humans came from meteorites.

But early metalworkers soon learned that iron could also

be found in the earth. By heating iron ore to a very high

temperature, they could separate the iron from the rock.

This process is called smelting. But the smelted iron was

not very strong and therefore not very useful. ☀ That

changed in about 1500 B.C. A Middle Eastern tribe called

the Hittites learned that by repeatedly hammering, heating, and cooling smelted iron, it would become very strong. They had learned to make steel. ☀ For many years, the

A crater created by a fallen meteorite

Hittites kept their steel-making methods secret, but valuable

secrets have a way of getting out. By 500 B.C., people all over

Europe and Asia were smelting iron and turning **The most common form of iron ore is hematite, a red rock containing 70 percent iron.**

it into steel. This was the beginning of the Iron

Age. ❋ Back then, no one understood how

iron became steel. We now know that adding

a small amount of carbon to iron produces steel.

The smelted iron picked up tiny amounts of carbon from

the charcoal fires used to heat the furnaces. Reheating and

Molten iron extracted through smelting

Early iron cookware

hammering the hot iron drove out the impurities and gave the steel a very strong, hard surface. ☀ One of the first uses for iron was for weapons. Steel knives and swords were stronger, sharper, and more deadly than the old bronze weapons. Iron was also used for cookware, plow blades, nails, and tools.

The Mysterious Force

In ancient times, people knew that certain kinds of rocks had a mysterious attraction to each other. They called these

Magnetite, a type of iron ore

rocks lodestones. Lodestones contain magnetite, a type of iron ore. When a lodestone hangs from a string, it always turns the same side toward the north—much like a compass. Travelers carried a lodestone with them to help find their way.

A World of Iron

Today, iron is our most common and useful metal. Millions of tons of iron are mined every year from great open-pit mines. The ore is smelted in giant **blast furnaces** to extract the iron. Most iron is then combined with carbon to

Steel bridges are strong and durable

make steel. ☼ Iron is used in everything from paper clips to

nuts and bolts to skyscrapers. Cars, bridges, **The red color of blood comes from iron, which carries oxygen through the body.**

eating utensils, and thousands of other

things are made mostly of iron and iron

alloys. Today, 3,000 years after the Hittites

first turned raw iron into steel, we are still in the Iron Age.

Exposed iron rusts, turning reddish-brown

Iron Oxidation

What You Need

Two jars

Two large steel nails

Water

One jar lid or plastic wrap

What You Do

1. Fill one jar to the top with water. Drop a steel nail into the jar and cover the jar with plastic wrap or a lid.

2. Fill the other jar 1/4 full with water. Drop a steel nail into the jar. Do not cover the jar.

3. Set the jars in a safe place and wait two days.

What You See

After two days, the nail in the open jar will have a heavy coat of rust. The nail in the covered jar will not be as rusty. To oxidize, or rust, iron needs both oxygen and water. The open jar lets plenty of oxygen get to the water, so the nail rusts quickly. The water in the sealed jar has only a small amount of oxygen, so the nail rusts slowly.

Rusty iron nails

Index

Words to Know

alloys (AL-oyz)—mixtures of two metals

blast furnace (BLAST FUR-nis)—a very hot furnace used for smelting iron and other metals

element (EL-e-ment)—a pure substance that cannot be broken down chemically; iron, sulfur, and oxygen are examples of elements

oxidation (AHX-i-DA-shun)—a process in which oxygen combines with another element to produce a new compound; when iron oxidizes, iron and oxygen combine to make the red iron oxide we call rust

smelting (SMELL-ting)—melting ores to extract metal

stainless steel (STAYN-lus STEEL)—an iron-nickel alloy that is resistant to rust

Read More

Bates, Robert. *Mineral Resources A-Z*. Hillside, N.J.: Enslow, 1991.

Fitzgerald, Karen. *The Story of Iron*. Danbury, Conn.: Franklin Watts, 1997.

Knapp, Brian. *Iron, Chromium, and Manganese*. Danbury, Conn.: Grolier Educational, 1996.

Internet Sites

Ask-a-Geologist
http://walrus.wr.usgs.gov/docs/
ask-a-ge.html/

Ironworld
http://www.ironworld.com/

The National Mining Hall of Fame and Museum
http://www.leadville.com/miningmuseum

INFORMATION